The Danvers Butchery
Meat Market and Cold Storage
Danvers, Massachusetts

Investigated by: John R. Anderson

This is Report 151 of the Major Fires Investigation Project conducted by Varley-Campbell and Associates, Inc./TriData Corporation under contract EME-97-CO-0506 to the United States Fire Administration, Federal Emergency Management Agency.

Department of Homeland Security
United States Fire Administration
National Fire Data Center

U.S. Fire Administration Fire Investigations Program

The U.S. Fire Administration develops reports on selected major fires throughout the country. The fires usually involve multiple deaths or a large loss of property. But the primary criterion for deciding to do a report is whether it will result in significant "lessons learned." In some cases these lessons bring to light new knowledge about fire--the effect of building construction or contents, human behavior in fire, etc. In other cases, the lessons are not new but are serious enough to highlight once again, with yet another fire tragedy report. In some cases, special reports are developed to discuss events, drills, or new technologies which are of interest to the fire service.

The reports are sent to fire magazines and are distributed at National and Regional fire meetings. The International Association of Fire Chiefs assists the USFA in disseminating the findings throughout the fire service. On a continuing basis the reports are available on request from the USFA; announcements of their availability are published widely in fire journals and newsletters.

This body of work provides detailed information on the nature of the fire problem for policymakers who must decide on allocations of resources between fire and other pressing problems, and within the fire service to improve codes and code enforcement, training, public fire education, building technology, and other related areas.

The Fire Administration, which has no regulatory authority, sends an experienced fire investigator into a community after a major incident only after having conferred with the local fire authorities to insure that the assistance and presence of the USFA would be supportive and would in no way interfere with any review of the incident they are themselves conducting. The intent is not to arrive during the event or even immediately after, but rather after the dust settles, so that a complete and objective review of all the important aspects of the incident can be made. Local authorities review the USFA's report while it is in draft. The USFA investigator or team is available to local authorities should they wish to request technical assistance for their own investigation.

This report and its recommendations was developed by USFA staff and by Varley-Campbell and Associations, Incorporated (Miami and Chicago), its staff and consultants, who are under contract to assist the Fire Administration in carrying out the Fire Reports Program.

The United States Fire Administration greatly appreciates the cooperation received from Danvers Fire Department.

For additional copies of this report write to the United States Fire Administration, 16825 South Seton Avenue, Emmitsburg, Maryland 21727. The report and the photographs, in color, are available on the Administration's Web site at http://www.usfa.dhs.gov.

U.S. Fire Administration

Mission Statement

As an entity of the Department of Homeland Security, the mission of the USFA is to reduce life and economic losses due to fire and related emergencies, through leadership, advocacy, coordination, and support. We serve the Nation independently, in coordination with other Federal agencies, and in partnership with fire protection and emergency service communities. With a commitment to excellence, we provide public education, training, technology, and data initiatives.

TABLE OF CONTENTS

The Danvers Butchery
Meat Market and Cold Storage
August 13, 2002

Investigated By: John R. Anderson

Local Contacts: Lieutenant Dave R. DeLuca
                   Fire Investigator
                   Danvers Fire Department
                   64 High Street
                   Danvers, MA 01923

OVERVIEW

In the early morning hours of August 13, 2002, Danvers, Massachusetts fire personnel were on the scene of a stubborn fire in a 30 yard dumpster packed with construction debris. Meanwhile, an arsonist allegedly set a fire outside The Danvers Butchery one mile away. The ignition location was both behind a fence and blocked from any neighbor's view by a tree line. The time of day, approximately 03:00 hours, also made it unlikely that anyone would notice.

The fire spread on the exterior vinyl siding and wood sheathing quickly engulfing the entire vertical surface. Soon, flames entered the interior of the 100 year old barn. The fire was reported nearly simultaneously by a neighbor and the alarm company which detected an interior fire. A reduced first alarm assignment arrived two minutes later and encountered a well involved structure fire. An aggressive attack with large hand lines knocked the fire down, and an interior attack was initiated. Due to structural concerns, inside efforts were terminated, and an exterior attack was resumed.

Suppression crews prevented flames from entering connected parts of the complex, but smoke and water damage was significant. Although the structural loss was high, the aggressive attack by the Danvers Fire Department saved a large volume of inventory as well as a large percentage of the equipment and fixtures within the building.

KEY ISSUES

Issues	Comments
Delayed Fire Reporting	The fire was allegedly set by an arsonist in the middle of the night on the building's exterior where no detection devices were present. The fire had to enter the structure before being detected. A neighbor also noticed the fire, but it was not reported until the detection system alarmed.
Lack of Fire Rated Barriers	Once inside the building, the fire had few barriers to slow its spread. Only the rapid response and aggressive attack by the fire department saved the complex. The original building was expanded on several occasions, but no barriers separated the additional compartments.
Unsprinklered Commercial Space	The age of this complex made it exempt from fire sprinkler requirements, although their presence would have reduced the fire loss.
Large Hose Streams Used for Quick Knockdown	Faced with a large fire volume, firefighters attacked with a 2-1/2" hose line from a hydrant supplied engine and a 1-3/4" preconnect from a quint's onboard water supply.
Predetermined Mutual Aid Plan	A total of six engines and one ladder company responded to Danvers during the fire. The engines closest to the fire were dispatched there to assist, and one established a crucial second water supply.
Fireground Hazards Were Present	The complex's electrical service lines disconnected from the building and were arcing against the ground in a key suppression location.

FIRE DEPARTMENT

The Town of Danvers, Massachusetts was incorporated in 1757 and abuts the historic city of Salem. It sits approximately 17 miles north of Boston, and has experienced tremendous growth over the past 20 years like the rest of the North Shore region. The community has 25,212 permanent residents but experiences a daily influx of as many as 50,000 people who work in the many businesses, shop in an abundance of retail stores, or attend schools. Many more simply travel through town on Interstate 128, Interstate 95, or U.S. Route 1. There is a mix of residential, commercial, and industrial occupancies, and the town also hosts North Shore Community College, Essex Agricultural Technical Institute, and St. John's Preparatory School. The land area protected is 14.09 square miles.

The Danvers Fire Department is charged with protecting this diverse community with 51 career personnel. The department is led by a Chief and Deputy Chief. A single Lieutenant performs fire prevention and fire investigation functions. The remaining personnel are divided into 12-man shifts who work a 42-hour week with two 10-hour days, two 14-hour nights, and four days off. Each shift is commanded by a Captain, but a Lieutenant may fill in during a vacation or sick day. Fire Headquarters is located in the downtown area and houses Engine 1, a 100-foot, 2,000-gallons-per-minute (gpm) quint with an assignment of four firefighters; Rescue 1, a mid-sized rescue truck with extrication, search, and medical equipment and carrying the shift officer and a firefighter; and Engine 3, a conventional 1,250-gpm pumper with a Lieutenant and two firefighters. Although the shift officer generally rides the Rescue, he has the discretion of going on any truck depending on the call. Also housed at this station are a reserve engine, a forestry truck, an Essex County owned pick-up truck with a foam trailer and a Decontamination Trailer which is part of the Massachusetts's hazardous materials response system.

Station 2 is located 3 miles away near the old Danvers State Hospital which is now closed. Engine 2 is a 75-foot, 1,250-gpm quint with a Lieutenant and two firefighters. Also in the station is a county owned aerial truck. All engines have 500 gallon water tanks except for Engine 2 which carries 400 gallons. Standard shift staffing is 12, and no less than 10 personnel are on a shift. An absent officer is often replaced by a senior firefighter. Engines 2 and 3 always maintain 3-man crews while Engine 1 may operate with as few as two. The Rescue has a 2-man assignment and routinely operates with Engine 1 as a two piece company.

New appointees must complete the 11-week training program at the Massachusetts Firefighting Academy. All members are Massachusetts's Emergency Medical Technicians. Firefighters are equipped with NFPA approved turnout gear and Scott 4.5 SCBA's. Training, vehicle maintenance, and other specific roles are performed by designated officers in addition to their supervisory and fire suppression responsibilities.

The department utilizes a 483-megahertz (MHz) radio system and can communicate with area agencies via multiple channel portable and vehicle radios. Dispatch is performed by civilians located at the Police Department. Two dispatchers are present at all times with one handling police calls and the other dealing with fire and emergency management services (EMS). During crisis situations, they assist each other to properly manage communications. A second alarm fire results in the call back of the lead dispatcher. Danvers has a long standing EMS contract with a local ambulance firm which provides advanced life support (ALS) and basic life support services, and are automatically dispatched to working fires.

BUILDING HISTORY AND CONSTRUCTION

The Danvers Butchery dates back to the early 1900s when the original building was constructed as a typical New England agricultural barn. It does not appear on the Danvers Town Map of 1897, but is referenced in the early part of the 20th century. The original "Barn" was of post and beam construction and measured 36 feet wide by 53 feet deep (approximately 1,908 square feet). Although the building is often referred to as 10 Donegal Lane, its official street number is 11. Many official records still have it listed improperly.

Over a period of years, several single story additions were constructed resulting in a complex with 9,385 square feet of retail, storage, kitchen, and office space. The basic footprint of the complex has not changed since 1979. (See Appendix A for site plan and Appendix B for floor plan) The complex had 147 feet of frontage, 18,000 square feet of paved parking, and a land area of .729 acres or 31,750 square feet. In fiscal year 2003, the property was assessed at $518,100. Of this, $285,300 was for the building, $211,100 was for the land, and $21,700 was for the paving.

The Barn had 2-inch by 4-inch walls throughout with exterior sheathing of 1-inch boards. The exterior finish was vinyl siding over tar paper, and the interior was 1/2-inch gypsum board. The vinyl siding was nailed directly to the sheathing with no added insulation. Additions were primarily wood frame construction, but concrete block was used at the east end. This was part of a renovation and replaced wood framed walls insulated with sawdust during early cold storage use. Prior to their replacement, the sawdust filled walls would expand and contract depending on the moisture content of the sawdust. They had lost their structural stability over time. During another renovation, the entire first floor east wall of the Barn was opened up to what became the retail store. A heavy timber beam and columns supported the Barn's east wall.

The ceilings in the retail and office areas were suspended. In some areas, the drywall extended just slightly above the ceiling channel, and open joist spaces were present for the convenience of running wires, pipes and communication cables. The original ceiling of the Barn's first floor was wood and was still present in many areas. Much of the Barn's second floor had open rafters to the attic and was only separated by the hung ceiling.

The complex sat on a series of concrete slabs. Slight differences in elevation demanded the construction of concrete ramps to allow for easy movements of products. There were no basements or crawl spaces. The roof of the barn was asphalt and the rest of the complex had a mix of asphalt and rolled roofing depending on pitch. Only the Barn's roof was affected by the fire.

The complex had a three phase electrical service to power the numerous cooling compressors. The supply was a pole with three can transformers located 40 feet southwest of the Barn's C/D corner. Cables ran to a drip loop and continued down via conduit on the C side of the Barn. The electrical room was in the B/C corner of the Barn on the first floor. The conduit and service entrance were directly above the area of origin.

BUILDING USE

The Barn as originally constructed had an obvious agricultural use. Early photographs show the typical large doors, hay loft door and hoist mechanism for lifting up the bales. During its life, one of the additions housed a bottling plant for a nearby fresh water spring, and others were used for butchering fresh killed game and storing the meat. In 1945, a business was opened on the site, Danvers Cold Storage Locker Plant, Inc. Area residents could rent one of the 297 lockers to preserve their meat. In 1950, 1,216 hunters used their services and paid $2.50 to skin a deer and 7 cents per pound for processing. The owner would also get the hides and tan them for manufacturing. He sold the goods at the Deerskin Trading Post across town. At a later point in time, a retail seafood store occupied the premises.

The property was purchased by the father of the current owner in the late 1970s and sold to the son around 1990. They developed the current business, The Danvers Butchery, which provided retail sales of meat. Some traditional groceries were also available for sale. Cooking and food preparation were done in a commercial kitchen located at the west end of the complex. The east end of the complex held several large walk-in freezer units to store perishable stock.

At the time it was built, the surrounding area was mostly farm land. It is currently a residential neighborhood and zoned as such. The Butchery is grandfathered for its use due to the longevity of the business. A typical year would see over 100,000 customers with an average of nearly 320 per day. Saturdays were the busiest with counts close to a thousand during the warm summer months when New England grills are fired up. While the store was not open on Sundays in consideration of the neighborhood, Saturdays frequently saw cars parked all along the roads in this area. Some neighbors disliked the business and kept a vigilant eye on the property. Numerous complaints were filed with the Danvers Building Department anytime changes were made at the property. Consequently, permits were reviewed regularly with appropriate inspections. A second store is located in nearby Newbury.

BUILDING FIRE PROTECTION

A combination burglar and fire alarm system was installed and operating within the building at the time of the fire. Monitoring was by a central station through a dedicated telephone line, and a back-up radio transmitter was located in the attic of the Barn. The burglar alarm system utilized a combination of motion detectors and door monitors. The fire alarm system consisted of ceiling mounted heat detectors and smoke detectors. There were two motion activated lights on the C side of the complex, but their unreliability dictated an "always on" status after the business closed for the day. An external fire alarm horn was located on the A side of the Barn, but it is unclear if it functioned. If power was interrupted prior to its activation, it may not have alerted neighbors. Residents of the house across the street from the Barn were not alerted to call the fire department.

Prior to the fire, the last recorded alarm activity was on Sunday morning, August 11th, when the business manager opened the door at 09:03 and secured the premises at 10:04. While the Butchery was closed, he often used the quiet time to complete purchasing for the next week. There were no sprinkler or other suppression systems in the complex with the exception of the kitchen where the hood and ducts were protected.

Municipal Water Supply

The Town of Danvers has a very good municipal water system fed by two reservoirs. Each has a capacity of over 5 million gallons and sit at an elevation of 234 feet above sea level. This allows the majority of the water system to be gravity fed, and four inline fire pumps assist the sections of town lacking pressure. The system is maintained by the Department of Public Works' (DPW's) Water Department which works closely with the Fire Department. Most hydrants are opened and flushed annually and repairs are made as needed.

The hydrant on Donegal Lane used during this incident was gravity fed and had a static pressure of 80 pounds per square inch (psi), a residual pressure of 75 psi, and a fire flow rate of 3,905 gpm at 20 psi. Several years ago, fire department personnel went around town and noted hydrant locations. This data was then input into the dispatch computer system. Approaching companies can request hydrant locations from dispatch when needed.

Building Codes

There were no known building codes pertaining to the construction of a barn in the early 1900s. The single story additions are not recorded and may have preceded the codes enforced today. Later renovations, however, would have required building permits and inspections, and many are documented at the Danvers Building Department. The Barn, in particular, saw many different uses and all were grandfathered under Massachusetts' Law. No renovations were extensive enough to bring the complex up to current standards.

THE FIRE

On the morning of August 12, 2002, the Danvers Fire Department was operating at a minimum manning level with only 10 firefighters on duty. At 02:44 hours, dispatch put out a first alarm for a report of flames coming from a building at 14 School Street. The assignment consisted of Engine 1, a 100-foot quint with two firefighters; Engine 3, a pumper with a senior firefighter as officer and two firefighters; and Rescue 1 with a Lieutenant and one firefighter. The Lieutenant was filling in as shift commander for the vacationing Captain.

Upon arrival three minutes later, they found a 30-yard construction dumpster well involved. It was near a structure, but flames were contained within the dumpster.

A 1-3/4-inch hand line from Engine 1 was initially used to attack the fire, but the dumpster was packed with debris making extinguishment difficult. The waste company which owned the container was notified at 03:05, and a roll-off truck was requested to move the dumpster to a location where it could be emptied. While waiting, firefighters secured a hydrant, committed the quint, and charged the ladder pipe which was elevated over the fire to soak its contents. The fire was of suspicious origin.

At approximately 03:00 hours, a fire started on the exterior of The Danvers Butchery. The area of origin was in a fenced enclosure next to a large air conditioning unit on the B/C corner of the Barn. (See Appendix C for details) The material first ignited is believed to have been a plastic bag filled with garbage. Sufficient combustibles were present to allow the fire to grow. Flames ignited the vinyl siding, tar paper, and sheathing. In several places flames ate through the sheathing and entered joist spaces. As flames impinged on the second floor window above, the glass failed. The fire entered the second floor and traveled through the open attic rafters once the suspended ceiling collapsed. It became visible to an awakened neighbor when the wood sheathing and vinyl siding on the C side were free burning. The property was shielded by a tree line which may have blocked earlier observations if any neighbors had, in fact, heard anything.

At 03:28 hours, the alarm company for the Danvers Butchery received a fire alarm for the attic. At 03:29 another fire alarm on the second floor initiated. It was a smoke detector at the top of the stairs outside Office G. (See Appendix B for 2nd floor plan) Seconds later, the motion detector in Office G activated. The alarm company notified the Fire Department and the building owners. The motion detector could have activated due to falling debris or a short circuit from the heat.

At 03:30 hours, dispatch received two nearly simultaneous reports of a fire at the Danvers Butchery at "10" Donegal Lane. The alarm company reported an activated fire alarm and a resident at 1 Elerton Lane reported flames showing. Engine 2 was dispatched at 03:31 as a first floor fire alarm activated at the Butchery and Engine 3 cleared the dumpster fire at School Street. The shift commander, C13, ordered Engine 3 to report conditions upon arrival, but a Danvers Police Sergeant arrived in the interim and reported a building heavily involved. C13 boarded Rescue 1 and responded with one other firefighter. Engine 1 remained at the dumpster with two men. The two fires were approximately one mile apart.

Engine 3 arrived at 03:33 and took a position on the A side of the building. The truck parked across the street at a hydrant, and the driver connected a short length of 4-inch hose. Engine 2 got to the scene less than a minute later, passed Engine 3 and proceeded to the east driveway entrance. The acting officer and a firefighter from Engine 3 began pulling a 2-1/2-inch line down the west driveway. As Rescue 1 turned from Route 35 onto Donegal Lane, C13 observed the intense fire conditions and requested a second alarm. Flames were coming from the A side window at the attic level and had breached the entire roof ridge of the Barn. The entire exterior C wall was burning and the fire had already consumed the roof gable on this side. (See Appendix D for photographs) C13, now the incident commander, knew the establishment well and realized no one would be in the building at this hour. Consequently, no interior searches were done. He was also familiar with the layout and contents.

Mutual Aid was started on the second alarm. Middleton and Topsfield each sent an engine to the fire. Danvers Headquarters station was covered by a ladder from Beverly and engines from Salem and Wenham. A call back of Danvers firefighters was begun, and 12 off-duty firefighters came in to assist. The Chief, Deputy, and Fire Inspector/Investigator were all notified to respond to the scene. The lead dispatcher was called to report to the dispatch center.

Engine 2, a 75-foot quint with a Lieutenant and two firefighters, did not pass a hydrant and did not lay in a supply line. They took a position on the C side of the complex about 50 feet east of the Barn. Normal procedure would have dictated this action, but, unlike most fires, no apparatus was immediately available to lay in a supply. The building's electrical service line had broken away from the drip loop and was arcing in the driveway outside the C/D corner. The electric company was notified, but the presence of the live wires precluded setting up a supply line from Engine 3 to the quint.

Engine 3's 2-1/2-inch was charged and directed at the attic and upper section of the D side. Because of the live wire, Engine 3 could not pass the C/D corner but was able to wash down the C wall from the D side. A 1-3/4-inch preconnect was pulled from Engine 2 and charged with tank water to attack the C side and to prevent extension into the abutting compressor shed. The initial suppression effort was being addressed by eight firefighters.

The Incident Commander walked around the complex to assess conditions. When he got to the rear, he observed Engine 2 without a supply and running low on tank water. Mutual aid companies were still five or ten minutes away since neither Topsfield nor Middleton has personnel on duty at night. Anticipating an extended exterior attack, C13 ordered Engine 2 to lay in their own supply line to the nearest hydrant and prepare for ladder pipe operations. Engine 2 shut down and disconnected the hand line. The crew wrapped a hydrant in front of 1 Elerton Lane and left one firefighter to charge the hose. About 500 feet of 4" supply line was laid back to the fire, and the quint was positioned on the C side once again. While opening the hydrant, which was found to be defective, the stem broke off rendering it useless. (See Appendix E for Fire Ground Overview)

Engine 2 had no water supply, and an incoming mutual aid company was ordered to continue the supply line to the next hydrant. During this time period, a unit radioed Danvers Dispatch and requested a hydrant location. Dispatch had no additional hydrants in their database.

Suppression with the 2-1/2-inch was very effective. The fire on the C wall was minimized, and the attic fire was knocked down. Firefighters observed flames inside the Barn's first floor at the C/D corner. Rather than continuing the exterior attack and pushing the fire further into the building, an interior attack was started. The store's front door (D1) was forced open, and two firefighters advanced a 1-3/4-inch line into the structure from Engine 3. Smoke and heat conditions were severe and visibility was less than a foot. The crew fought their way about 80 feet to the C side and extinguished all fire found on the first floor. During their entry time they radioed Command with reports of their progress.

The Department Chief, C1, arrived and was briefed by C13. He ordered C13 to remain in command and provided assistance. The interior crew reported fire rolling above them but out of sight. C13 knew the fire was still burning on the second and attic levels and had observed several sections of the roof sliding off the rafters. Since the ridge beam was also weakened, he ordered everyone out of the building and initiated an exterior attack.

A mutual aid Engine passed the broken hydrant and proceeded down Elerton Lane in search of a new water source. The only other hydrant is located where the road terminates at a cul-de-sac. They laid 500 feet of 4-inch supply line into Engine 2's hose at the dead hydrant and secured a water source for Engine 2. While a water supply was now present, it was not run through a pump, and the resulting pressure and flows were insufficient to sustain an effective master stream from Engine 2's ladder pipe. The total length of the lay was over 1,000 feet. Engine 2 did flow water through their pipe, and they also kept a 1-3/4-inch in use. Due to Engine 2's position and the orientation of the Barn's roof, they could only access the east side for ventilation or suppression.

The 2-1/2-inch from Engine 3 remained active on the C and D sides of the Barn. Engine 3 utilized a deck gun from its position on the A side to complete the master stream attack. The 1-3/4-inch hose used for the interior attack was now used from the exterior. Engine 3's 1,250-gpm pump, supplied by a strong hydrant, had no trouble keeping up with the flow. This effort was maintained for about 1-1/2 hours. Danvers's reserve, Engine 4, was staffed by off duty personnel and responded. Engine 1 was relieved at School St. by Salem and Wenham engines and proceeded to Donegal Lane for aerial operations on the west side of the Barn roof. Hamilton and North Reading sent engines to cover Danvers Headquarters.

The State Fire Marshal's Office was notified at 04:29 hours, and a State Trooper assigned to the fire arrived at 05:14. The fire was regarded as suspicious. When the majority of the visible fire was suppressed, hand lines were once again deployed to complete the task. A Positive Pressure Ventilator was set up at the front door, and crews entered the building to overhaul at about 05:30 hours. Multiple sections of the roof were opened to access pockets of fire that could not be safely reached from the interior.

The overhaul became a significant activity due to the complexity of the building. The Barn's use for cold storage in the mid 1900s had resulted in a mixture of insulation materials. Firefighters had to open many walls to search out smoldering materials which included hay stuffed into joist spaces. The Department's thermal imaging camera was used to detect hot spots during overhaul. The Building Inspector and Health Inspector were both called in at 06:13. All food products in the retail portion of the store had to be disposed of. At 06:35 mutual aid companies at the scene were released, and at 07:21 Danvers's crews began rotating out to bring in the fresh day shift.

One or two pieces of apparatus were at the scene for the next five hours to complete overhaul and wet down hot spots. At 14:00, a section of the roof collapsed. The building inspector had previously ordered that the remaining portion be demolished and removed once suppression was complete. Firefighters returned to the scene once in the afternoon at 15:42 to do a final wet down.

In the end, only the Barn sustained actual fire damage, but the entire complex experienced significant smoke damage. Even in the Barn, the extent of fire damage was varied. Rooms on the second floor that lacked a resistant barrier to the attic, a partial wood floor above the suspended ceiling, were heavily damaged while others had only moderate heat, smoke, and water damage. Only the rear portion of the first floor had fire damage.

The calculated loss was about $800,000, but the aggressive exterior and interior attacks by the Danvers Fire Department saved not only two thirds of the building, but many of the coolers, display cases, and kitchen equipment. Additionally, over $40,000 in stored meat products were salvaged from the coolers at the east end since their heavy doors resisted smoke penetration. These refrigeration units alone were worth over $15,000. Circumstances would have been much different if the fire had consumed more of the first floor. There were no injuries to civilians or firefighters.

Recall was sounded at 16:24 hours. Investigators determined that the cause was arson, and a suspect was arrested. He is currently awaiting trial.

LESSONS LEARNED

1. **The arsonist remains a difficult challenge to fire prevention.**

   Most fire codes are written to control conditions within a building, but outside arson fires have plagued communities for years. While it is probably beyond the scope of the local fire inspector to do much about this, there are some preventative measures that building owners can take.

   First, the presence of lighting will deter some crimes, and while motion detector activation is not highly reliable, it may be preferred by some over the cost of constant illumination. It was policy to have yard lighting on at the Butchery, but it is unknown if they were working. Power was already disconnected when the first trained eyes got to the scene.

   Second, the material believed to have been used to start the fire was removed from a dumpster on the premises. If the dumpster was chained shut, the garbage bag would not have been available. Motive and opportunity may not have changed, but the ignition material would not have been available.

   Third, good housekeeping around a building will eliminate many potential combustibles which may be used for illegal purposes or might hinder fire department access. The obvious community benefit can't be ignored either.

   It would have been useful if the property had been surrounded with security fencing.

   All four of these suggestions could be offered during a routine fire inspection, and would be seen as proactive.

2. **Delayed reporting resulted in a well developed fire requiring a second alarm.**

   The outside point of origin is clearly the major factor in the growth of the fire without detection. It was well advanced when a neighbor was alerted, and the alarm system only functioned once the fire entered the structure. Earlier detection would have likely required substantially less suppression efforts and resources.

   The possible application of external detection devices should be explored for high risk property. In some western areas, primarily with common dry ground cover and serious wildfire risks, external fire sprinklers are currently used to protect buildings from exposure. Adapting wildfire technologies to urban structures may provide another loss prevention approach, especially in confined areas such as alleyways.

3. **Commercial buildings should be designed to limit the spread of flame and smoke between sections.**

   Because of the high fire risk, protective devices like fire doors were historically incorporated into commercial and industrial buildings. In today's structures, they have largely been replaced by fire sprinklers which are designed to limit or extinguish flames and for life safety. A sprinkler system does not manage smoke. With increasing amounts of petroleum based materials in buildings and their voluminous smoke characteristics, smoke travel deserves more attention.

   The building damage beyond the physical "Barn" was largely caused by smoke, and the resulting cleaning, repair, or replacement added to the high fire loss. The smoke spread was made possible by the removal of the Barn's first floor, east wall prior to the fire. Compartmentalization could have prevented a lot of damage. A fully sprinklered building, including the attic, would have substantially limited the interior destruction.

4.   Structures venting flame upon the fire department's arrival call for large hose streams.

The acting officer of the first arriving fire apparatus, Engine 3, observed a structure fire venting flames at the attic level and through the roof and made a quick decision about the suppression method. With only three firefighters, the truck position was textbook since it did not block the road and allowed the driver to hook up the hydrant without assistance.

Often, a fire can be quickly knocked down using large hand lines or even master streams, and traditional interior attacks can follow. Any attack initiated with large water volumes must be supported by an adequate water supply, and conventional fire trucks seldom carry sufficient water to sustain large water flows for any extended period. It must also be realized that this technique is being used to knock down, and usually not extinguish, heavy fire. Water must be available to complete suppression and continue through overhaul.

The lack of a water supply for Engine 2 limited the crew's effectiveness during the Butchery fire. The combination of a broken hydrant, lack of a close back-up hydrant, and a long supply line without pump support made this a difficult tactical objective. Fortunately, the effectiveness of Engine 3's deck gun and hand lines limited the urgency of that challenge.

5.   Some buildings should be brought closer to code compliance.

While The Danvers's Butchery is hardly unique in being "grandfathered" for years, questions remain for such buildings about the extent that upgrades can and should be enforced. The economic implications of improvements cannot be ignored, but basic construction practices like rated horizontal barriers should be required. Renovation plans for the Barn in 1990 show 1 hour rate stairway enclosures and doors, and yet an open stairway to the attic remained. The second floors suspended ceiling merely covered up open rafters in many places including Office G where the fire quickly spread into the attic.

While it may be cheaper for the business to pay more for insurance than to do improvements, the building and firefighters' safety suffer. If it had been possible to enter the attic during this event, a firefighter could have easily fallen through open rafters that were not detectable from below.

The fire caused enough damage that the repairs would mandate code compliance throughout the complex, which is probably not cost effective. Current plans are to raze the complex, sell the land for two or three house lots, and build a new store in a different location that offers more parking and less controversy with neighbors.

6.   A predetermined mutual aid plan initiated early during a serious fire helps assure adequate personnel and equipment.

The Danvers Fire Department had a mutual aid and call back plan in place prior to this fire. Consequently, adequate personnel were present to assist at the scene and cover the community. In all, seven area communities provided help with six engines and one ladder truck.

An appropriately equipped engine also provided an important second water supply. The early activation of mutual aid got this supply in place a lot quicker than other strategies would have allowed. Command would have had to wait for recalled firefighters to respond with Engine 4 since Engine 2 was committed to a location without sufficient staffing to move the 4" line and attempt taking another hydrant.

Command's decision to fall back to an exterior attack did not necessitate use of all the mutual aid companies, but they were available if needed.

7. **The effectiveness of the fire service cannot always be judged by dollar loss.**

The Danvers Butchery fire had a relatively high dollar loss and yet much of the structure is intact or certainly repairable. Its use as a market left inventory exposed, and it had to be thrown out. Some display and cooling equipment in the center section of the complex was also lost. At the same time, a significant amount of the contents was saved including $40,000 in food and many coolers, freezers, display cases, ovens, and other equipment whose value is in the tens of thousands of dollars.

Danvers is a popular North Shore community with very high land and property values. The tax assessment for this property is over $500,000 with a land value well over $200,000. The Barn was less than half of the property's square footage and was only in average condition when last appraised. Its replacement value, however, is quite high at current construction costs. Since the entire complex was closed up after the fire without heat or electricity, the damage from a harsh winter on the structures and mechanical systems made the loss worse.

In reality, the aggressive suppression efforts of the Danvers Fire Department saved a substantial portion of property. The entire complex could have easily burned to the ground, but effective incident command, appropriate tactics, and a good water supply prevailed.

APPENDICES

Appendix A: Maps of Danvers Butchery

Appendix B: Map: Complex--First Floor

Appendix C: Map: Complex--First Floor Showing Area of Origin

Appendix D: Photographs

APPENDIX A

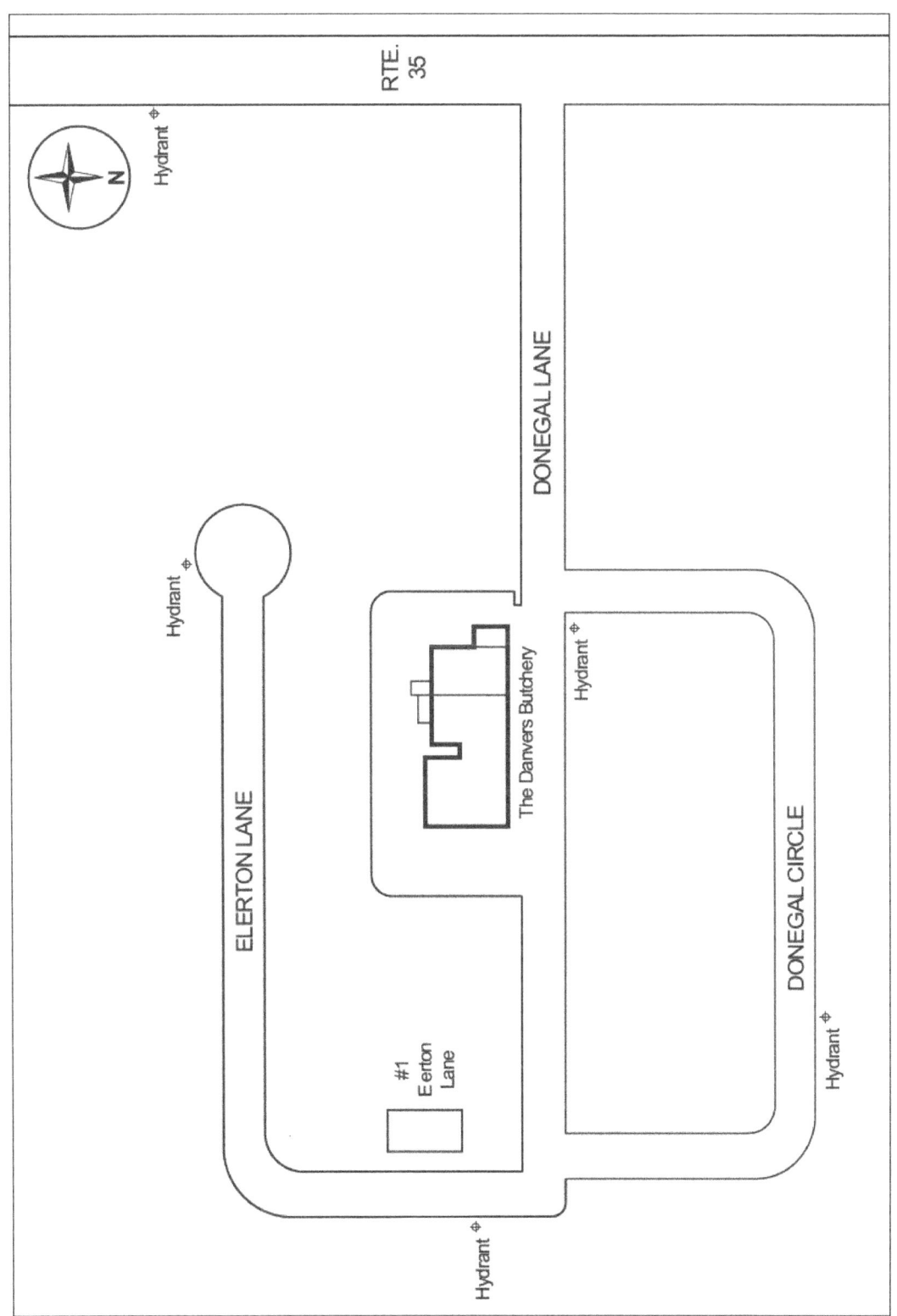

Map 1: Street Plan

Appendix A (continued)

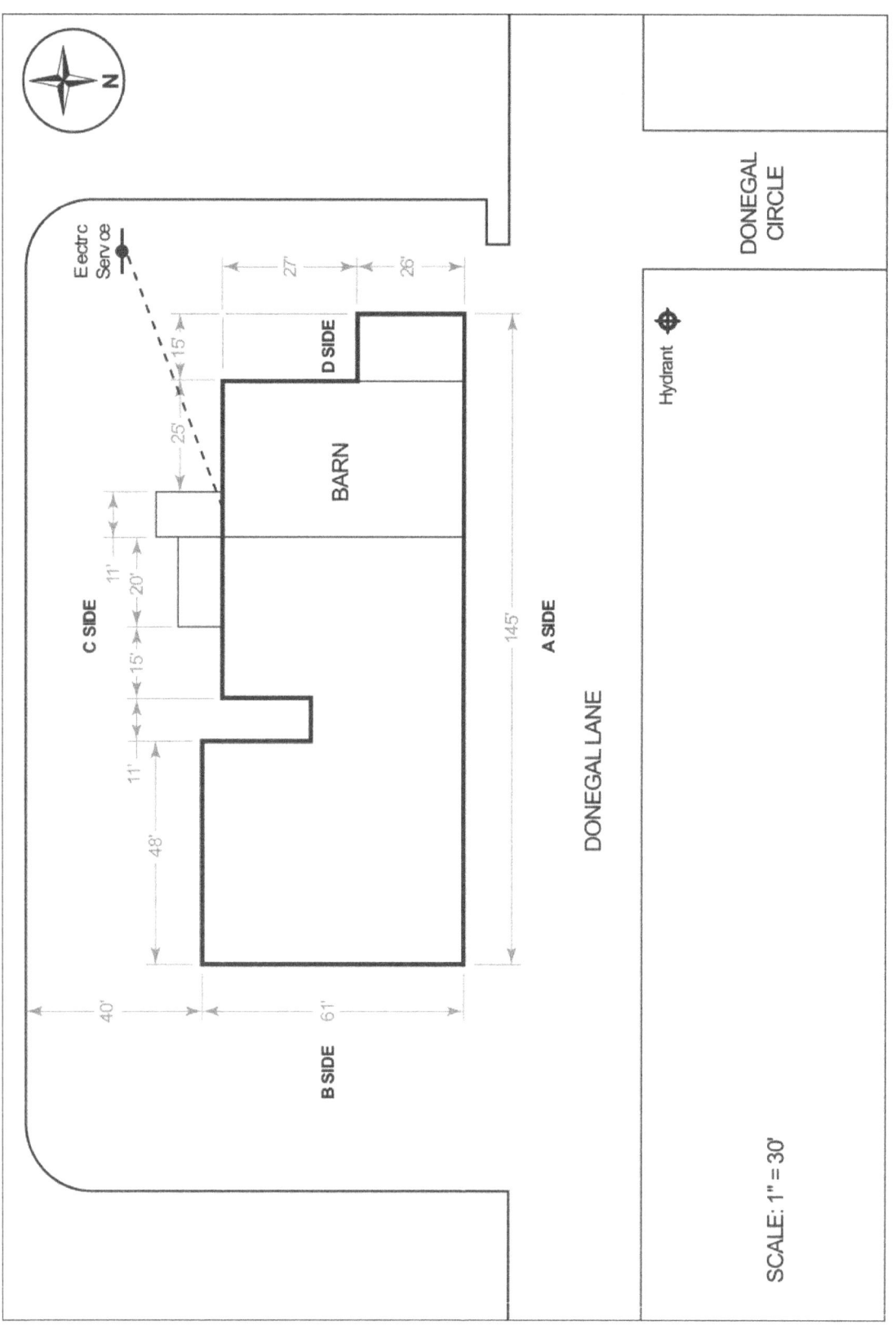

Map 2: Site Plan

Appendix A (continued)

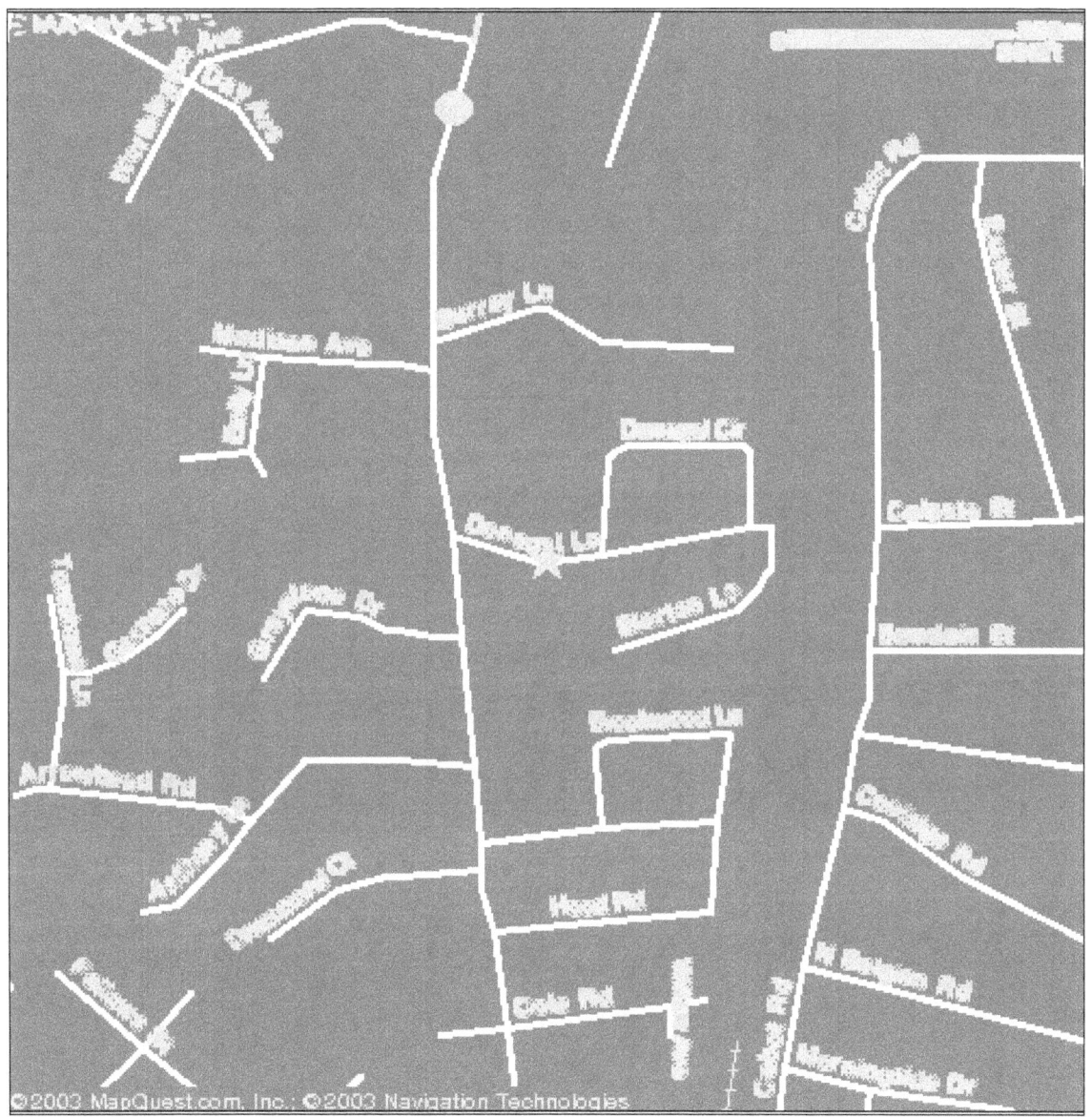

Map 3: Donegal Lane, Danvers, MA

Map: Complex – First Floor

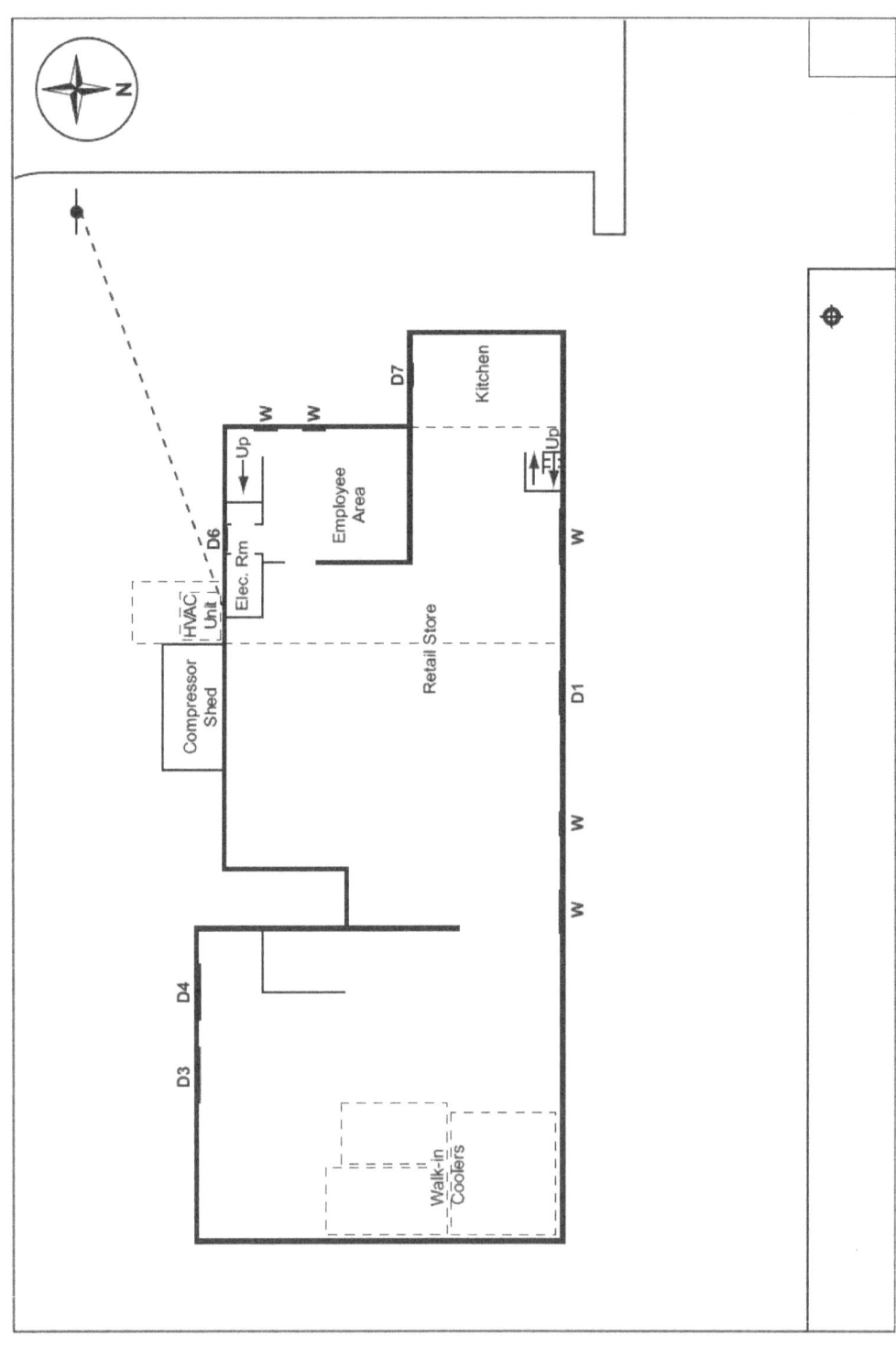

APPENDIX C

Map: Complex – First Floor

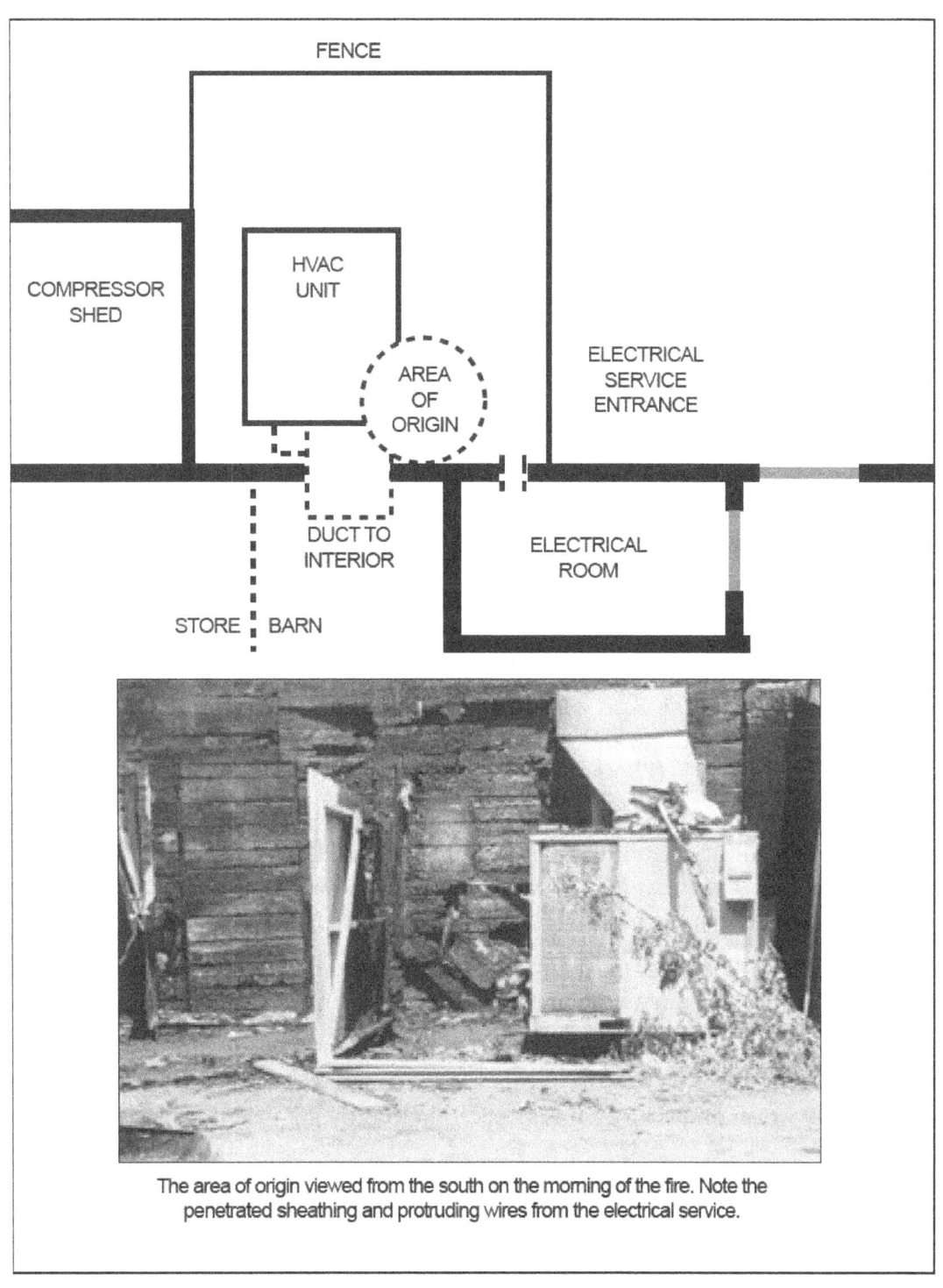

The area of origin viewed from the south on the morning of the fire. Note the penetrated sheathing and protruding wires from the electrical service.

APPENDIX D

1. The eastern portion of The Danvers Butchery's Front, A side. Danvers's Engine 4, the reserve, is still on the scene. The hydrant tagged by Engine 3 is in the foreground. The interior attack was through D1, the door in the center white section. Note that the glass is still present in many windows. (Photo courtesy Danvers Fire Department)

2. The western portion of the A side and the D side. Arriving firefighters observed flames from the front attic window and along the ridge. Danvers's Engine 1, a quint, is shown. (Photo courtesy Danvers Fire Department)

3. The upper part of the Barn's C side. The gable had burned through before firefighters arrived. The unstable roof was removed after suppression was completed. (Photo courtesy Danvers Fire Department)

4. The lower portion of the Barn's C side. This surface, finished with vinyl siding, was engulfed when firefighters arrived. The area of origin is at the lower right. (Photo courtesy Danvers Fire Department)

5. The rear of the Barn, 1st floor looking southwest. The protruding wall sections in the center are the outside of the electrical room,. The duct from the outside HVAC unit is at the left and the entrance to the employee area is through the door right of center. The beam at the top is holding up the Barn's east wall. (Photo is by the investigator)

6. The north part of the Barn, 1st floor looking west into the kitchen. The fire did not penetrate the interior wall shown on the left. This photo was taken about six months after the fire, and a lot of salvage work has been done. (Photo is by the investigator)

7   The employee area on the Barn's 1st floor, looking south at the D6. The electrical room is through the door on the left, and the door on the right accesses a closet under the rear stairway to the second floor. (Photo is by the investigator)

8. Another view of the employee area, looking southwest. The entrance to the staircase is at the end of the left wall where it intersects with the west wall, the C/D corner of the Barn. Fire damage in this area was caused by debris dropping down from above. The interior attack extinguished the fire in the room. (Photo is by the investigator)

9. The Barn's second floor, looking south from the conference room. This room sustained only heat and water damage due to the presence of a wood floor in the attic. Office G can be seen through the French doors. It was devastated. (Photo is by the investigator)

10. These are the walk-in coolers at the east end of the complex. The smoke damage was minimal, and $40,000 worth of inventory was saved. (Photo is by the investigator)

Appendix D (continued)

1. The eastern portion of The Danvers Butchery's Front, A side. Danvers's Engine 4, the reserve, is still on the scene. The hydrant tagged by Engine 3 is in the foreground. The interior attack was through D1, the door in the center white section. Note that the glass is still present in many windows.

Appendix D (continued)

2. The western portion of the A side and the D side. Arriving firefighters observed flames from the front attic window and along the ridge. Danvers's Engine 1, a quint, is shown.

Appendix D (continued)

3. The upper part of the Barn's C side. The gable had burned through before firefighters arrived. The unstable roof was removed after suppression was completed.

Appendix D (continued)

4. The lower portion of the Barn's C side. This surface, finished with vinyl siding, was engulfed when firefighters arrived. The area of origin is at the lower right.

Appendix D (continued)

5. The rear of the Barn, 1st floor looking southwest. The protruding wall sections in the center are the outside of the electrical room. The duct from the outside HVAC unit is at the left and the entrance to the employee area is through the door right of center. The beam at the top is holding up the Barn's east wall.

Appendix D (continued)

6. The north part of the Barn, 1st floor looking west into the kitchen. The fire did not penetrate the interior wall shown on the left. This photo was taken about six months after the fire, and a lot of salvage work has been done.

Appendix D (continued)

7. The employee area on the Barn's 1st floor, looking south at the D6. The electrical room is through the door on the left, and the door on the right accesses a closet under the rear stairway to the second floor.

Appendix D (continued)

8. Another view of the employee area, looking southwest. The entrance to the staircase is at the end of the left wall where it intersects with the west wall, the C/D corner of the Barn. Fire damage in this area was caused by debris dropping down from above. The interior attack extinguished the fire in the room.

Appendix D (continued)

9. The Barn's second floor, looking south from the conference room. This room sustained only heat and water damage due to the presence of a wood floor in the attic. Office G can be seen through the French doors. It was devastated.

Appendix D (continued)

10. These are the walk-in coolers at the east end of the complex The smoke damage was minimal, and $40,000 worth of inventory was saved.